Caveman Teach Children Metric System Systems International (SI)

Math and Science

Written and Illustrated
By
Rod Rammage

Copyright 2021

Caveman likes to teach children math and science. Children need to learn math and science. Math and science are good for children. Caveman will tell a story to teach children about the Metric System.

Many years ago, before children were born, Caveman lived in Kilo Village. Many people lived in Kilo Village. Everyone in Kilo Village worked hard to help each other.

One day, many people were working together to build a big hut for big parties. Village people like parties. The villagers had a big problem. No one was able to measure the same.

One man used his foot to measure. No one else's foot was as big as his.

One woman used her whole body to measure. Some people were short and some people were tall.

One villager even tried to measure with his nose. His nose was very dirty.

Everyone measured differently. The villagers were building a crooked building. Everyone was unhappy.

The villagers gave Caveman a job. The villagers wanted Caveman to invent a new system for measuring so everyone would measure the same.

Caveman needed time to think. Caveman went to a quiet place and sat on a rock. Caveman had a big dog named Meter Dog. Meter Dog sat down next to Caveman. Maybe Meter wanted to help Caveman think.

A big, long necked dinosaur named Hecto Dinosaur distracted Caveman. Hecto Dinosaur was eating leaves from the Kilo Village tree.

While Caveman and Meter were thinking, a mammoth named Deka Mammoth walked by. Deka Mammoth distracted Caveman by making a big sneeze.

A mouse named Deci Mouse came out and began playing in the dirt. Deci mouse made Caveman laugh.

Caveman watched a big, red ant crawl across the rock. Caveman named the big, red ant Centi Ant. Caveman remembered once getting bitten by Centi ant; it burned like fire.

Meter Dog started jumping around and biting himself. Deci Mouse got scared and ran away. Meter Dog jumped, bit, and scratched himself until a flea jumped off.

The flea juimped on a rock by Caveman. Caveman decided to name the flea Milli Flea. Milli Flea was very small. Meter Dog was afraid of Milli Flea. Caveman thought Milli Flea must be very itchy for Meter Dog to be afraid.

Caveman watched Meter Dog. Suddenly, Caveman jumped up. Meter Dog had given Caveman an idea for a new measurement system. Caveman decided to name the new system after Meter Dog. Caveman named the new system the Metric System. Caveman used all the animals he had seen for the new system.

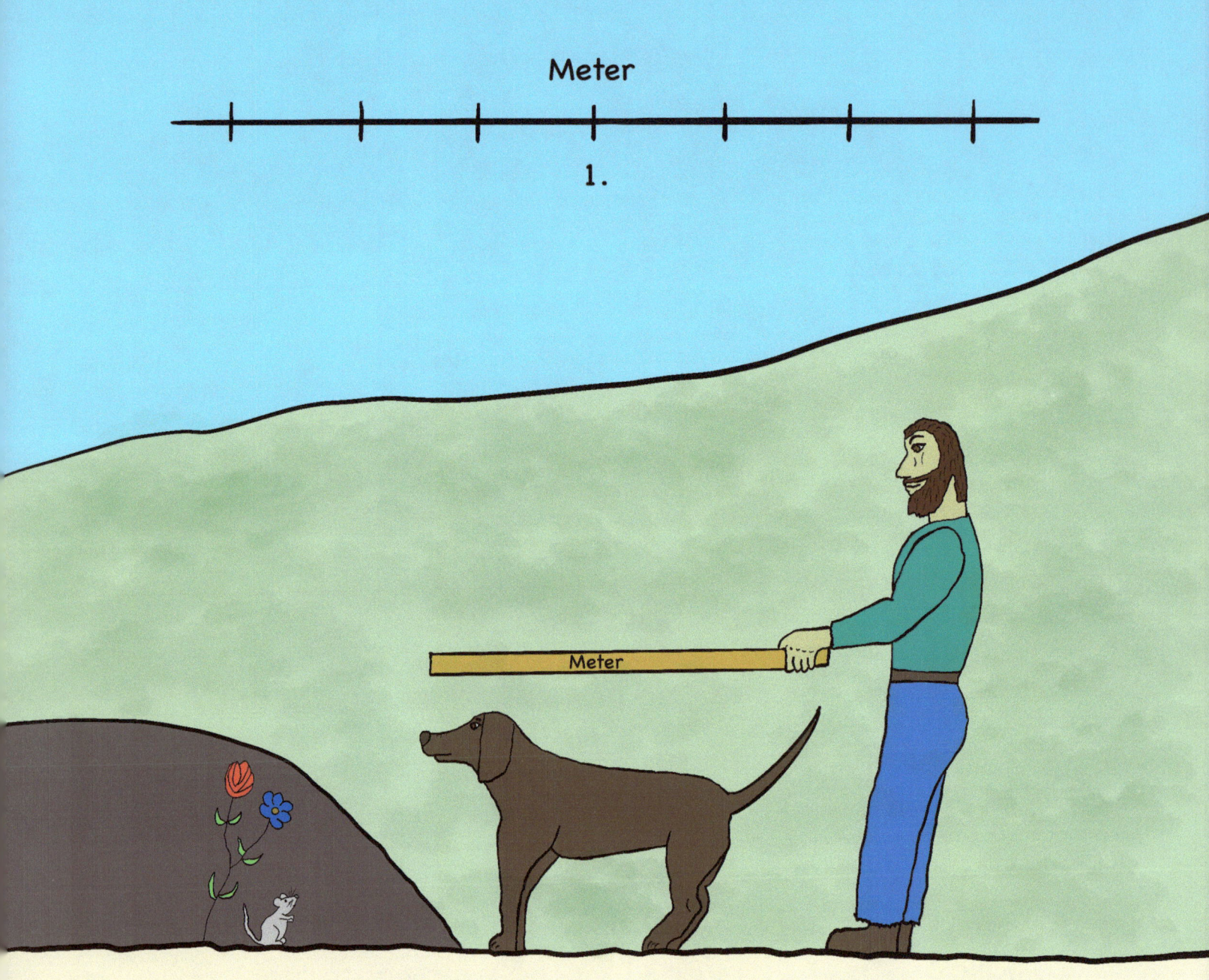

Caveman made a stick the same length as Meter Dog. Caveman called the stick a meter. From then on, all meter sticks were the same length as Meter Dog.

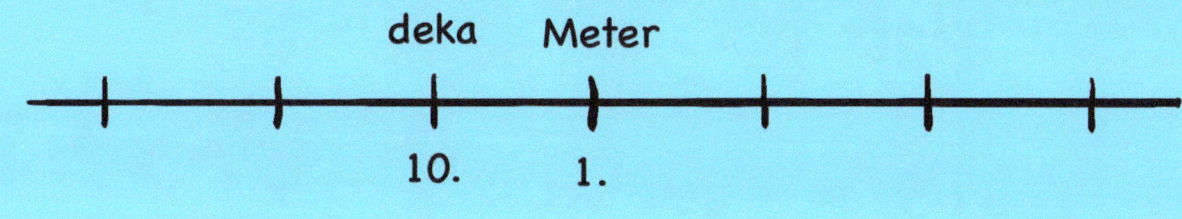

It took ten Meter Dogs to make one Deka Mammoth. Deka Mammoth was ten times longer than a meter. From then on, all dekameters were ten times as long as meters.

It took ten Deka Mammoths to make one Hecto Dinosaur. Hecto Dinosaur was ten times longer than a dekameter. Caveman called this distance a hectometer.

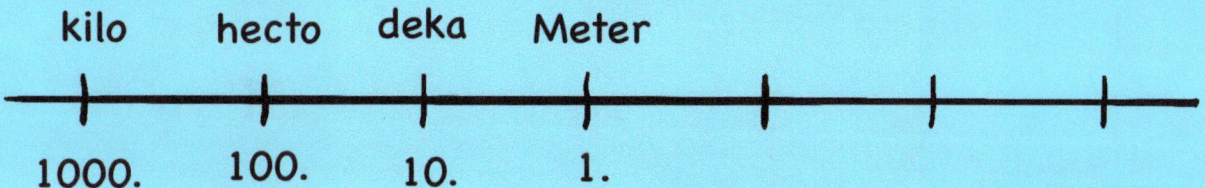

It took ten Hecto Dinosaurs to make one Kilo Village. Kilo Village was ten times longer than a hectometer. Caveman called this distance a kilometer.

It took ten Deci Mice to make one Meter Dog. Deci Mouse was ten times smaller than Meter Dog. That was Meter Dog divided by ten. Caveman called this distance a decimeter.

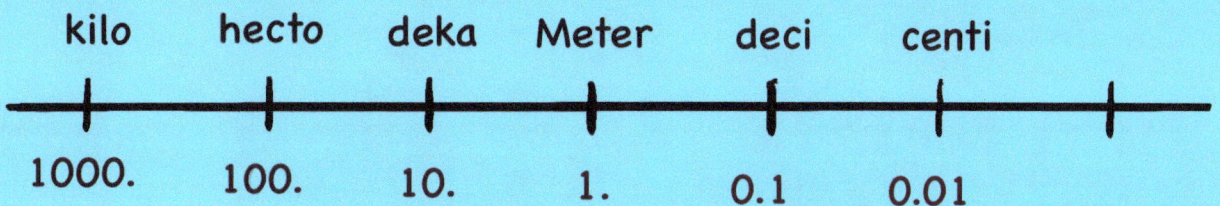

It took ten Centi Ants to make one Deci Mouse. Centi Ant was ten times smaller than Deci Mouse. That was Deci Mouse divided by ten. Caveman called this distance a centimeter.

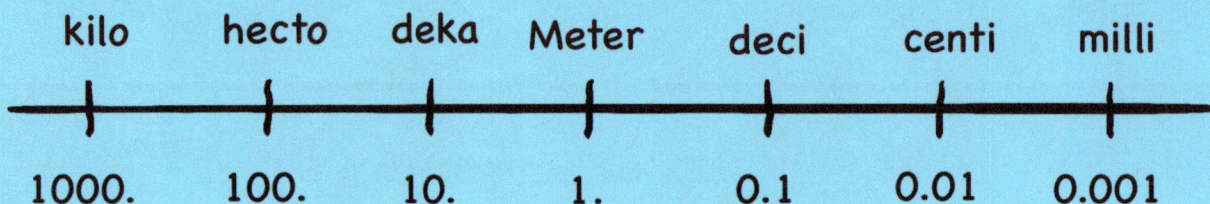

It took ten Milli Fleas to make one Centi Ant. Milli Flea was ten times smaller than Centi Ant. That was Centi Ant divided by ten. Caveman called this distance a millimeter.

Now everyone in the village was able to use the same measurement system and build straight buildings. Everyone was happy.

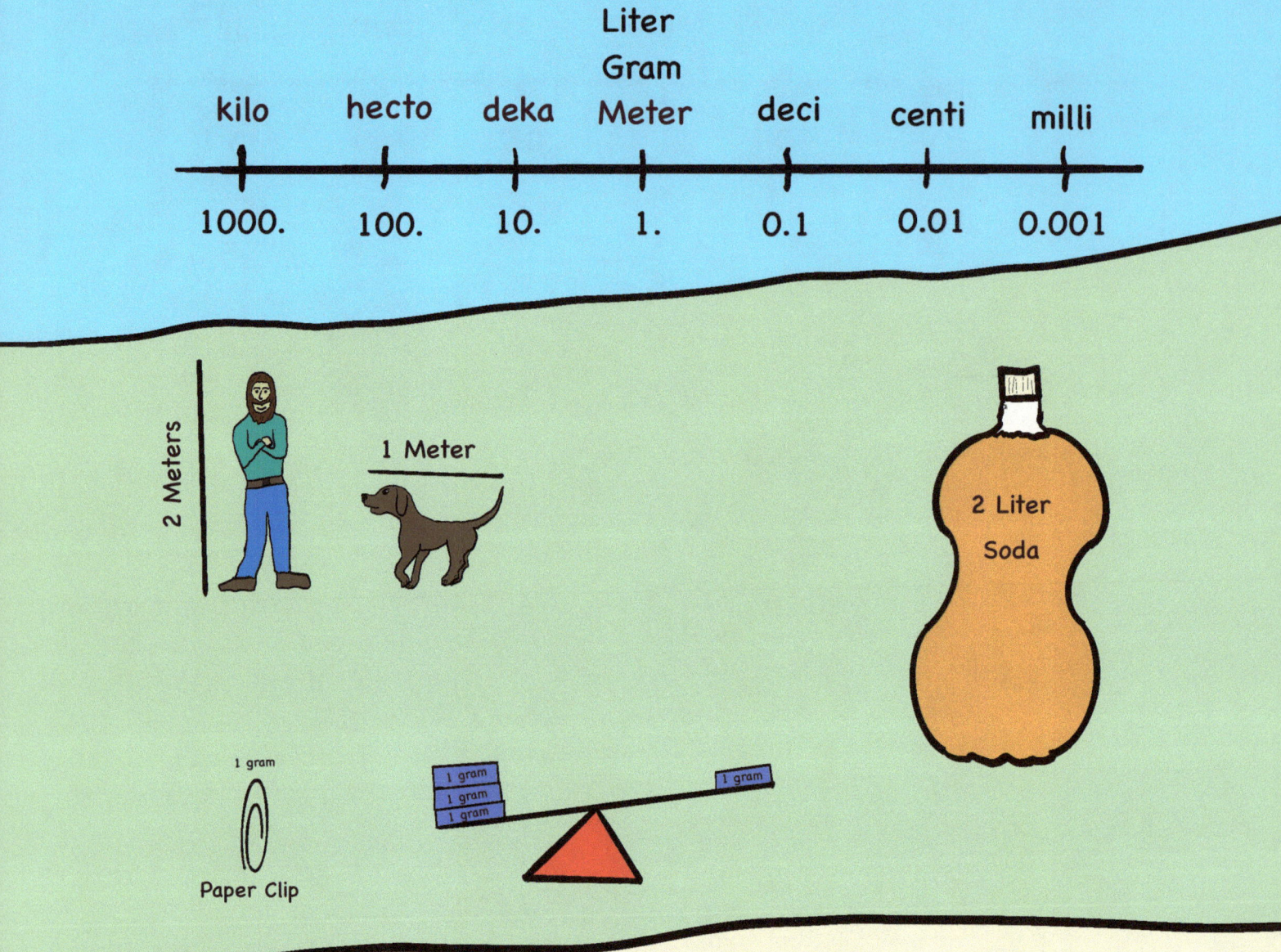

Caveman said the Metric System—also known as the Systems International (SI) units—could be used in three different ways. Meters would be used to measure distances. Grams would be used to measure weight (mass). Liters would be used to measure size (volume).

Caveman was glad Meter Dog gave him the idea for the Metric System. Caveman was itchy. Maybe Caveman had a Milli Flea.

Metric System Comprehension Questions

Beginner
1. How big is a meter?
2. How big is a dekameter?
3. How big is a hectometer?
4. How big is a kilometer?
5. How big is a decimeter?
6. How big is a centimeter?
7. How big is a millimeter?

Intermediate
8. What number does a meter represent?
9. What number does a dekameter represent?
10. What number does a hectometer represent?
11. What number does a kilometer represent?
12. What number does a decimeter represent?
13. What number does a centimeter represent?
14. What number does a millimeter represent?
15. What do meters measure?
16. What do grams measure?
17. What do liters measure?

Advanced
18. How many meters are in a kilometer?
19. How many centimeters are in a meter?
20. How many millimeters are in a meter?
21. How many millimeters are in a kilometer?
22. How many centimeters are in a kilometer?

Metric System Comprehension Question Answers

1. About the size of a big dog.
2. About the size of a mammoth.
3. About the size of a long-neck dinosaur.
4. About the size of a small village.
5. About the size of a mouse.
6. About the size of a big red ant.
7. About the size of a flea.

8. One
9. Ten
10. One-hundred
11. One-thousand
12. One-tenth
13. One-hundredth
14. One-thousandth
15. Distance
16. Weight or mass
17. Size or volume

18. 1000
19. 100
20. 1000
21. 1,000,000
22. 100,000

Thank you for your time and attention. I hope you enjoyed this book as much as I enjoyed creating it. Storytelling has always been my favorite way of teaching. I have many more ideas I will be working on in the future. Please help me become a better writer and help other people enjoy my work by reviewing this book on Amazon.

https://www.amazon.com/review/create-review?&asin=b0bc76jphx

Meet the author at amazon.com/author/rodrammage

Contact me personally at:
yellowpencilmathematics@yahoo.com